SPACE FIRSTS™

# SKYLAB
# The First American Space Station

Heather Feldman

The Rosen Publishing Group's
PowerKids Press™
New York

*For Mark, the greatest husband, father, and friend in the world*

Published in 2003 by The Rosen Publishing Group, Inc.

29 East 21st Street, New York, NY 10010


First Edition

Editor: Nancy MacDonell Smith
Book Design: Mike Donnellan

Photo Credits: Cover, pp. 4, 7, 8, 11, 12, 15, 16, 19, 20 © Photri Microstock, Inc.

Feldman, Heather.
Skylab : the first American space station / Heather Feldman.— 1st ed.
v. cm. — (Space firsts)
Includes bibliographical references and index.
Contents: What is Skylab? — Spacehouse — Skylab's first crew — More astronauts on Skylab — Setting up Skylab — The human body in space — Learning about the sun — The end of Skylab — What we learned from Skylab — The International Space Station.
ISBN 0-8239-6248-2 (library binding)
1. Skylab Program—Juvenile literature. 2. Space stations—United States—Juvenile literature. 3. Space sciences—Juvenile literature.
[1. Skylab Program. 2. Space stations. 3. Space sciences.] I. Title.
TL789.8.U6 F45 2003
629.44'2—dc21

2002000103

Manufactured in the United States of America

# Contents

# What Is Skylab?

*Skylab* was the first American space station. A space station is a large structure in space that **orbits** Earth. **Scientists** and **astronauts** live in a space station for a period of time, sometimes for as long as several months. They conduct **experiments**, make observations, and study space. *Skylab* was **launched** on May 14, 1973, from the Kennedy Space Center in Florida. The *Saturn V* rocket blasted off and carried *Skylab* into space. *Skylab* was used for 251 days and was visited by three different crews of astronauts. It orbited Earth at an **altitude** of 270 miles (434.5 km). Most airplanes rarely fly higher than 7 miles (11 km) above Earth's surface. *Skylab* traveled almost 40 times farther from Earth than an airplane travels.

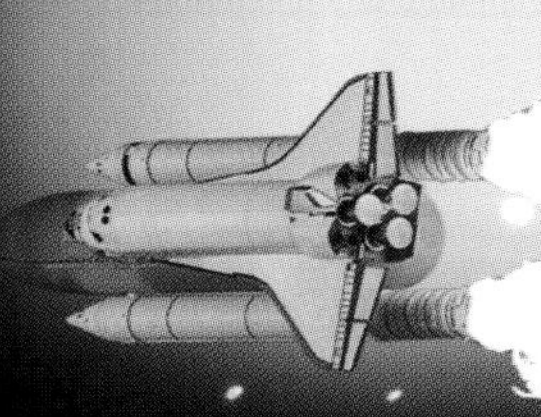

Skylab *was damaged during its first launch. This meant that the first crew had to wait more than a week before they could get on board the space station.*

# A House in Space

*Skylab* was divided into two levels and had as much room inside as a small three-bedroom house. *Skylab* weighed almost 100 tons (91 t)! In the spacehouse were living quarters, a small bathroom, a kitchen, and a workshop. The bedroom contained a sleeping bag for each astronaut, along with a light, headphones for music, and a curtain that could be closed for privacy. There were other parts to *Skylab*, as well. The dock allowed **space shuttles** to land on the station. The porthole was used for viewing Earth. One section of *Skylab* had telescopes used for observing the Sun and Earth. On the second floor, there was a large room used for exercising and playing games, such as catch. There was also something called an **airlock**.

*The airlock allowed the astronauts to go back and forth from the space station to space.* Inset: *One of the crew members uses* Skylab*'s shower.*

NASA
NASA
NASA

# Skylab's First Crew

*Skylab* was launched into space without a crew. The three-person crews were each sent up separately. The first crew arrived on *Skylab* on May 25, 1973. Several parts of the station had been **damaged** during the launch. The sunscreen/**meteorite** shield had been torn off. Astronauts Charles Conrad Jr., Paul Weitz, and Joseph Kerwin had to repair it. This protective screen was needed because high temperatures from the Sun could spoil the food and medical supplies and could make the inside of the station too hot to occupy. Fixing the sunscreen was a dangerous job. To protect *Skylab* from the Sun, the crew built a foil shield that looked like an umbrella. The repair was a success. *Skylab*'s first crew stayed for 28 days and got the station up and running. They orbited Earth 404 times and took 29,000 photographs!

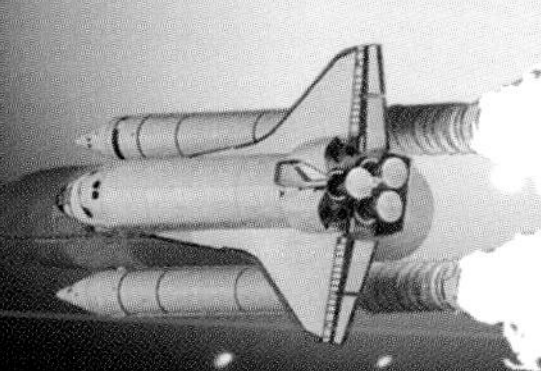

*Astronauts Kerwin and Conrad make repairs to* Skylab. Inset: *The first astronauts to live on board* Skylab *were, from left to right, Joseph Kerwin (science pilot), Charles Conrad Jr. (commander), and Paul J. Weitz (pilot).*

# More Astronauts on Skylab

The second crew of astronauts landed on *Skylab* on July 28, 1973. Alan Bean, Jack Lousma, and Owen Garriott spent 59 days on *Skylab* and did many scientific and medical experiments. They completed 858 orbits around Earth. The last crew to visit *Skylab* landed on November 16, 1974, and stayed for 84 days. Astronauts Gerald Carr, William Pogue, and Edward Gibson orbited Earth 1,214 times! This trip set a new record for the amount of time humans had spent in space. These last two groups of astronauts were sent to *Skylab* to prove that humans could live and work in space for long periods of time. Scientists wanted to find out what living in **zero gravity** did to the human body. Their other job was to study the Sun and how it produces **energy**. The three crews performed almost 300 experiments.

*The inside of* Skylab *was not much taller than an adult human.* Far right: *An astronaut from the third* Skylab *crew makes repairs to the space station.*

NASA

# Living on Skylab

*Skylab*'s first crew was in charge of setting up the spacehouse where the three groups would live. The crew had to unpack several months' worth of basic supplies and materials for experiments. At least 20,000 items were brought on board, including 2,000 pounds (907 kg) of food and 1,200 aspirin. Even though the astronauts had fixed the sunscreen, it was still warm in the station, and unpacking made it feel even hotter. However, the crew was surprised by how easy it was to handle boxes in a weightless **environment**.

The effects of **gravity** are not felt in space, so people and objects float around freely. To keep themselves from floating around *Skylab*, the astronauts wore special shoes that could be locked in place.

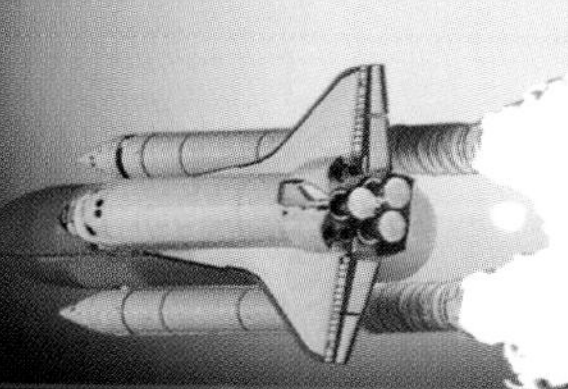

*The space station included a small kitchen. During meals the crew strapped down their feet and thighs so that they could stay seated at the table.*

# The Human Body in Space

All three crews that visited *Skylab* had to deal with living in a weightless environment for long periods of time. While on *Skylab*, they repeatedly took blood samples. When they returned to Earth, these samples would be studied to see what effects weightlessness could have on the body's ability to fight disease. Tests were also done to measure how well their hearts and muscles worked in a weightless setting. All three crews proved that humans could live in space for long periods of time. In fact, the longer the crews stayed in space, the more used to weightlessness their bodies became. The third crew seemed to bounce back to normal the quickest. All of the crews agreed that exercising in space frequently was a key factor in helping their bodies get used to life back on Earth.

*Astronaut Joseph P. Kerwin looks at Charles Conrad Jr.'s teeth. On Earth, Conrad would have had to sit in a dentist's chair. On Skylab, he could float upside down instead!*

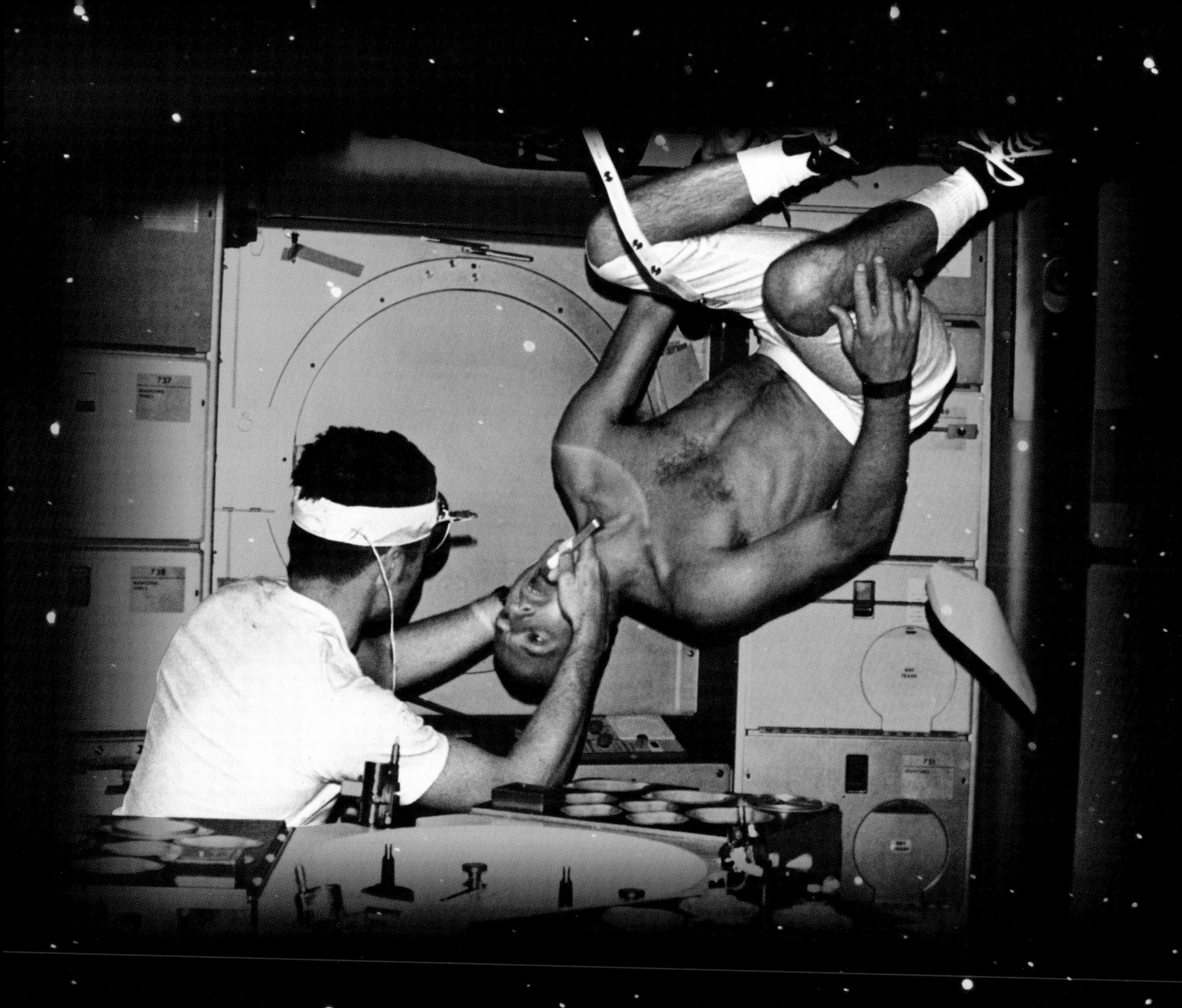

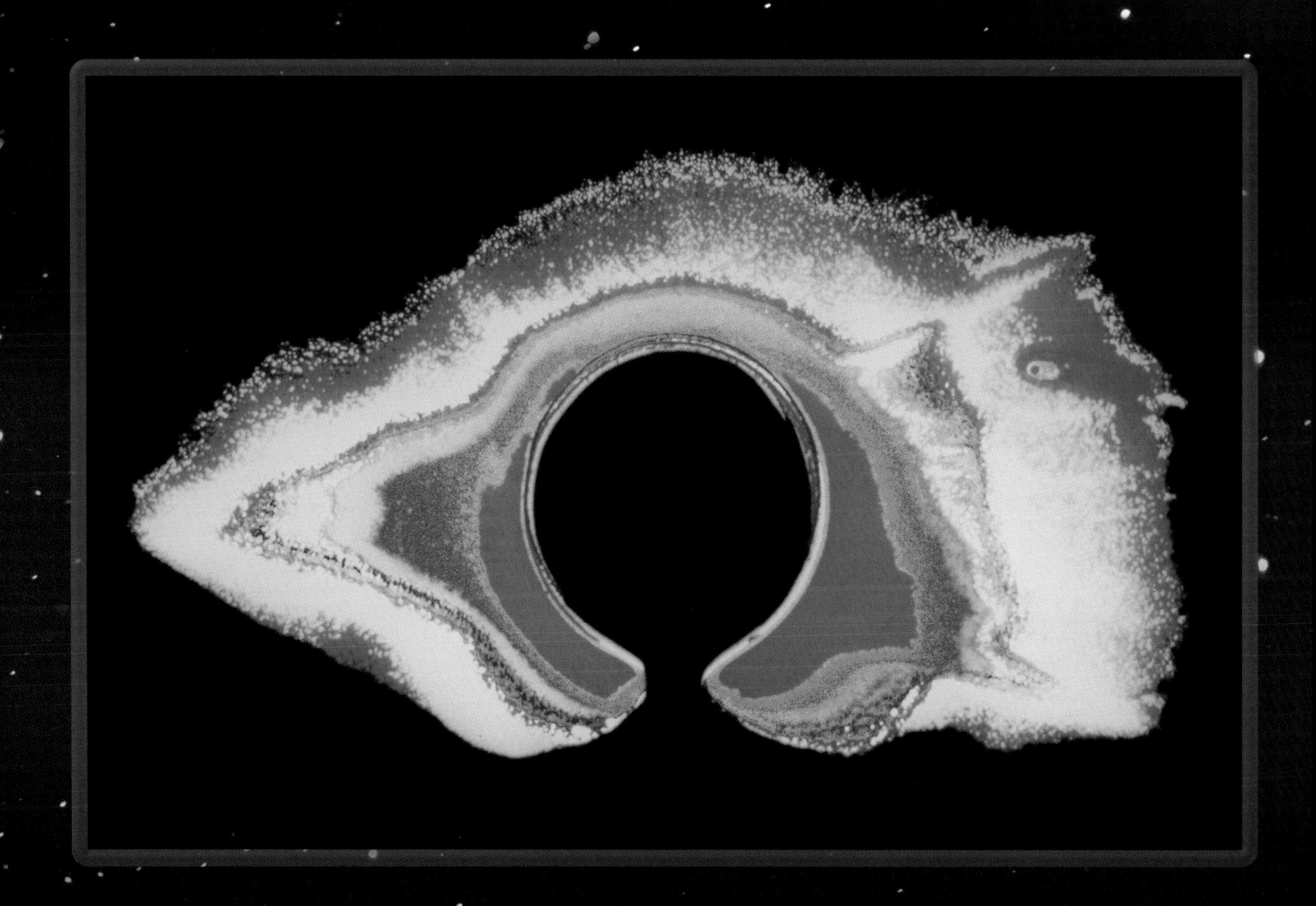

# Learning About the Sun

One of *Skylab*'s main goals was to research the closest star, the Sun. Scientists were able to conduct many experiments on *Skylab*. The three crews living on *Skylab* brought back a total of 175,000 pictures of the Sun. The crews on *Skylab* were the first to observe the part of the Sun called the **corona**. They began to learn of what this outer layer was made. For the first time, scientists could actually see **flares** occurring in the corona. These flares, or eruptions, on the Sun's outer layer give off so much energy that they can affect **radio waves** and cause power blackouts on Earth. They may also affect the weather patterns on Earth. The research gathered by *Skylab* crews provided scientists with many new details about the Sun.

*This photograph of the Sun's corona was taken by* Skylab*'s telescope.*

# The End of Skylab

The third and last group of astronauts left *Skylab* on February 8, 1974. However, scientists and **engineers** on the ground at the National Aeronautics and Space Administration (**NASA**) continued to run tests on *Skylab*. They performed some tests that would have been too risky had crews still been inside. For instance, on February 9, the engineers wiped out all of the instructions in *Skylab*'s computer, then tried to put them back in again. This test let scientists see what would happen if an emergency arose on *Skylab*. At 2:00 P.M., the systems were shut down, then were successfully turned on again. After these tests were completed, *Skylab* stayed in orbit for another five years.

*NASA scientists had planned to keep* Skylab *in orbit for 8 to 10 years, but the space station started to lose altitude in 1977, and fell out of orbit 2 years later. Pieces of* Skylab *landed in the Indian Ocean.*

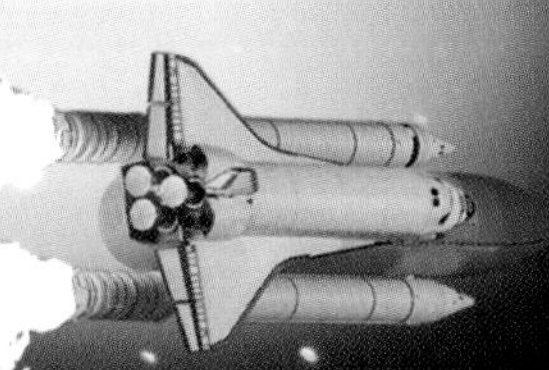

VIP

# What We Learned from Skylab

The *Skylab* project was a great success. America's first space station had two main goals. NASA scientists wanted to gain more knowledge about the Sun and prove that humans could live and work in space for long periods of time. *Skylab*'s crews accomplished both of these goals and more. *Skylab*'s photos were used to create detailed maps of various areas of the Sun and Earth. They were also used to record weather patterns. Studies were done on Earth's crops, soil, and oceans. Scientists were even able to map out good places to fish all the way from *Skylab*! The crews studied many things, all to improve life on Earth. *Skylab* provided new and exciting information for scientists. Most important, everyone returned home safe and healthy.

*Astronauts Charles Conrad Jr., Paul Weitz, and Joseph Kerwin, the first people to live on* Skylab*, return to Earth after completing their mission.*

# The International Space Station

*Skylab* led the way for the largest, most advanced space station in history. The *International Space Station* (*ISS*) was launched in 2000. The *ISS* is a research center that orbits Earth every 90 minutes. The United States, Russia, Canada, Japan, Italy, Brazil, and the European Space Agency are all partners in the *ISS*. This space station is the size of three average houses, and it weighs about 1 million pounds (453,592 kg)! The *ISS* provides an incredibly clear view of the world, so that humans can see and can solve some of the problems faced on Earth. It takes cooperation among people from many countries to make this "space hotel" work. Humankind will surely benefit from the amazing work being done by everyone involved in this historic project.

# Glossary

**airlock** (AYR-lahk) An airtight room permitting passage between spaces of different air pressure.
**altitude** (AL-tih-tood) Height above Earth's surface.
**astronauts** (AS-troh-nots) Members of a crew on a spacecraft.
**corona** (kuh-ROH-nah) The sixth, or top, layer of a star or the Sun.
**damaged** (DA-mijd) Harmed, broken.
**energy** (EH-nur-jee) The ability to do work.
**engineers** (en-jih-NEERZ) People who are experts at planning and building engines, machines, roads, bridges, and canals.
**environment** (en-VY-urn-ment) All the living things and conditions that make up a place.
**experiments** (ik-SPER-uh-ments) Tests done on something to learn more about it.
**flares** (FLAYRZ) Explosions on the Sun's surface that release a lot of energy.
**gravity** (GRA-vih-tee) The natural force that causes objects to move or to tend to move toward the center of Earth.
**launched** (LAWNCHD) When a spacecraft is pushed into the air.
**meteorite** (MEE-tee-uh-ryt) A rock that has reached Earth from outer space.
**NASA** (NA-suh) National Aeronautics and Space Administration, the United States's space agency.
**orbits** (OR-bits) Travels in a circular path around another body in space.
**radio waves** (RAY-dee-oh WAYVZ) Waves that are carried by changes in both electric and magnetic fields that include those waves used to carry radio and television signals.
**scientists** (SY-en-tists) People who study the world and the universe.
**space shuttles** (SPAYS SHUH-tulz) Reusable spacecrafts designed to travel to and from space carrying people and cargo.
**zero gravity** (ZEE-roh GRA-vih-tee) No gravity, no pull of Earth on any nearby body.

# Index



# Web Sites

Due to the changing nature of Internet links, PowerKids Press has developed an online list of Web sites related to the subject of this book. This site is updated regularly. Please use this link to access the list:
www.powerkidslinks.com/sf/skylab/